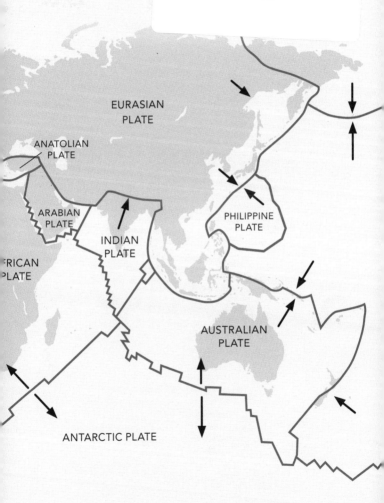

Series 117

This is a Ladybird Expert book, one of a series of titles for an adult readership. Written by some of the leading lights and outstanding communicators in their fields and published by one of the most trusted and well-loved names in books, the Ladybird Expert series provides clear, accessible and authoritative introductions, informed by expert opinion, to key subjects drawn from science, history and culture.

The Publisher would like to thank the following for the illustrative references for this book: Page 13: chart courtesy of the U. S. Geological Survey; page 25: map image from National Geographic Maps/ National Geographic Creative; page 29: reference for Fred Vine taken from photograph: Fred Vine and Drummond Matthews at the University of East Anglia, 1970 © The British Library by permission of Fred Vine; charts courtesy of the United States Geological Survey (USGS); page 37: diagram from A harbinger of plate tectonics: a commentary on Bullard, Everett and Smith (1965) 'The fit of the continents around the Atlantic', John F. Dewey, *Phil. Trans. R. Soc.* A 2015 373 20140227; DOI: 10.1098/rsta.2014.0227. Published 6 March 2015; page 41: top chart: 'Iapetus Fossil Evidence' courtesy of Wouldloper via Wikimedia Commons; bottom map: Permit Number CP17/078 British Geological Survey © NERC 2017. All rights reserved.

Every effort has been made to ensure images are correctly attributed; however, if any omission or error has been made please notify the Publisher for correction in future editions.

MICHAEL JOSEPH

UK | USA | Canada | Ireland | Australia
India | New Zealand | South Africa

Michael Joseph is part of the Penguin Random House group of companies whose addresses can be found at global.penguinrandomhouse.com

First published 2018
001

Text copyright © Iain Stewart, 2018

All images copyright © Ladybird Books Ltd, 2018

The moral right of the author has been asserted

Printed in Italy by L.E.G.O. S.p.A.

A CIP catalogue record for this book is available from the British Library

ISBN: 978-0-718-18718-7

www.greenpenguin.co.uk

Penguin Random House is committed to a sustainable future for our business, our readers and our planet. This book is made from Forest Stewardship Council® certified paper.

Plate Tectonics

Iain Stewart

with illustrations by
Ruth Palmer

Ladybird Books Ltd, London

Earth puzzle

Our earth is a puzzle – the product of a planetary experiment that was run once only and lasted four and a half billion years. Unravelling that succession of worlds which has unfolded until the present would seem to be a task of impossible complexity. For centuries, scientific enquiry had grappled with convoluted ideas to explain the earth's natural features: its continents and oceans, mountains and valleys, volcanoes and earthquakes.

Then, over the course of just a few years in the 1960s – a geological instant – it was all over. A small group of geophysicists – some leaders in the field, many just out of graduate school – working at a handful of international institutions, formulated the first global theory ever to be generally accepted in the entire history of earth science. Like all grand unifying theories, it would turn out to be deceptively simple.

Today, half a century on, it is hard to imagine thinking about our planet without the lens of plate tectonics. The theory not only sits at the heart of our scientific understanding of how the earth works, it has infiltrated into everyday life. People talk of the 'moving plates' of politics or 'tectonic shifts' in world affairs. But few other than earth scientists appreciate the deep roots and revolutionary currents that led to what would be one of the greatest scientific breakthroughs of the twentieth century.

This is the story of the science and scientists that made earth history.

Voyages of discovery

Throughout the sixteenth century, successive waves of Portuguese and Spanish seafarers charted the outer reaches of the known world. These grand voyages of discovery revealed an earth that was more ocean than land, had distinct climate zones and wind patterns, and exhibited a geomagnetic field.

Back home, European map-makers converted these maritime charts into ever more intricate and accurate world views. One, the renowned Flemish geographer Abraham Ortelius, noticed a curious transatlantic twinning. In his *Thesaurus Geographicus* (1596), he noted that western Europe and Africa seemed to project into the recesses of the American continent. Brazil's eastern bulge seemed to fit snugly into the bight of West Africa, the Sahara's western bulk could rest against North America's eastern shores, and Canada's Nova Scotia and Newfoundland appeared to slot into the Bay of Biscay and the English Channel.

By the nineteenth century, speculation about the origin of the earth's enigmatic natural features was the pastime of gentleman scientists and scholarly clergymen. Scientific theories were generally thinly veiled biblical thinking, often invoking divine violence. In 1838, the Reverend Thomas Dick, a Scottish minister and philosopher, wrote that the pairing of Atlantic shores 'renders it not altogether impossible that the continents were originally conjoined and that, at some favourable physical revolution or catastrophe, they may have been rent asunder by some tremendous power, when the waters of the ocean rushed in between them and left them separated as we now behold them'.

Abraham Ortelius with a later depiction of his idea.

Greater glory

March 1912: Robert Falcon Scott and his team are returning from their ill-fated attempt to be the first to the South Pole. Blighted by frostbite, snow blindness and malnutrition, and man-hauling 16 kg of rock samples, they still take time to 'geologize' the Transantarctic Mountains. Ultimately, it would cost them their lives. Almost eight months later, when their frozen bodies were found, Scott's rock samples were carefully laid out. He had clearly considered them precious cargo.

Scott had been persuaded to collect the rocks by a young English palaeobotanist, Marie Stopes. Later in life she would find fame as a pioneer of women's rights and family planning, but in the early 1900s she was an expert on ferns and seed plants from Carboniferous times, 300 million years ago. Certain that such rocks would be found in Antarctica, she had pleaded with Scott to take her on his expedition. He refused, but promised to bring her back some samples.

When analysed back in England, Scott's samples were found to be packed full of fossilized plant debris. Many of the leaves were from a fern-like tree called *Glossopteris*, indicating that in the Carboniferous age, this icy wasteland had been carpeted with temperate forest. The spores of these trees couldn't be transported great distances but *Glossopteris* had been found in Carboniferous strata right across the southern hemisphere, from India to South America. All those land masses must once have been welded together, including, now, Antarctica. Scott's rocks had helped to define the southern extent of an extraordinary land mass.

The face of the earth

Earlier, in the 1850s, the Vienna-based geologist Eduard Suess had discovered an ocean lurking amid the magnificent Austrian Alps. Finding thick piles of ancient sediment identical to that accumulating on the modern sea floor, he proposed that a primeval marine basin the size of the Atlantic had once dominated the heart of Europe. Suess named it Tethys, after the sister and consort of the Greek god of the ocean, Oceanos. Suess's crumpled Alpine peaks were testimony to slow constriction of the Tethys Ocean by the northward advance of a great land mass to the south.

In his 1904 book *The Face of the Earth*, a monumental work compiled over two decades, Suess forensically reconstructed the geology of the southern land mass to embrace South America, Africa, India and Australia. Its heart lay in the Gondwana kingdom of north central India, where *Glossopteris* fern forest had once flourished, and so the vast bulk became Gondwanaland. Opposing it on the northern shores of the Tethys was Angaraland, and as the Tethys Ocean squeezed shut, parts of the two super-continents became welded into the present amalgam of Europe and Asia.

Suess, the towering geological intellect of the day, explained the demise of the Tethys by likening the earth to a drying apple. As it contracted through loss of internal heat, its rocky skin wrinkled, pushing up mountains and sinking ocean basins. Oceans were simply sunken continents. In the death throes of the Tethys, land bridges interconnecting Gondwanaland successively foundered, giving birth to the Atlantic and Indian oceans and leaving the southern continents stranded. Continents were considered to be on the move, but mainly up and down.

A man adrift

Alfred Wegener was not a geologist. Famed as a world-record-setting balloonist, trained as an astronomer and working on atmospheric physics, the German meteorologist spent much of his time pioneering dangerous scientific expeditions to the polar north. Having lived for months on the shifting fringes of the Greenland ice cap, Wegener was ready to challenge the notion of a static planet.

In 1912, he combined Suess's grand synthesis with the latest geophysical thinking to make an outrageous proposal: the continents were drifting like ice floes. Over the next decade or so, Wegener showed a continuity of fossils and strata indicating dispersed lands and glacial deposits in regions that are ice-free today. By the time his book *The Origin of Continents and Oceans* appeared in English in 1924 he had drawn all the continents together 300 million years ago into a single ancestral land mass: Pangaea ('all earth').

In Britain and America, Wegener's 'continental drift' met with a frigid reception. His claim that Greenland was moving by tens of metres each year was ridiculed by leading physicists. What could possibly propel it at such a pace? For geologists, the problem wasn't so much the driving force, or even the evidence, but the irritation of a German physicist with his head in the clouds intruding rudely into their rocky domain. Held firm by a contracting earth dogma, most geologists found Wegener's idea simply too outlandish to accept. If drift was real, they complained, it would require a revolution in geology. His theory of continental mobility shredded by the scientific establishment, Wegener returned to Greenland and died on the ice in November 1930, searching for proof to the end.

A man of convection

On the opposite side of the Atlantic, even as Wegener and his 'drift' were being consigned to the history books, a viable mechanism for moving the continents was being conceived. The discovery of natural radioactivity at the end of the nineteenth century had allowed the planet a new heat source. Earth wasn't steadily cooling and contracting – it had an internal fuel store of radioactively decaying elements.

In the early decades of the twentieth century, the English geologist Arthur Holmes used the new knowledge that radioactive atoms shed their energy particles like clockwork to build a timescale of the planet's geological past. But in the late 1920s, he began to realize that the heat of radioactive decay could also power a mighty engine within the earth.

Holmes presented his theory of 'mantle convection' in a talk to the Geological Society of Glasgow in 1928. It appeared in print in 1931, just months after Wegener's death. Ironically, Holmes's account of the heat engine borrowed much from meteorologists' descriptions of the turbulent motions of the atmosphere. Hot rock in the earth's mantle (the bit between the crust and the core) could move as 'a planetary circulation' of 'monsoon-like currents' – albeit rising and falling only a few centimetres a year – stoked by simple thermal convection.

Arthur Holmes went on to be one of Britain's most influential geologists and his 1944 textbook, *The Principles of Physical Geology* – written during long hours on wartime fire-watching duty – inspired generations of geologists. Its final diagram depicts mantle convection, couched in cautious terms. Holmes might have been converted to drift, but few others had.

From land to sea

Despite the discovery of a potential mechanism for a mobile earth, most geologists doggedly clung to fixed continents left isolated by sunken land bridges. After all, the ocean floor was conveniently out of sight and largely out of mind.

Half a century before, in 1872, the HMS *Challenger* expedition had begun the first scientific study of the world's great oceans. Surveying for the very first network of deep-sea telegraph cables, soundings revealed a huge rise in the middle of the Atlantic. Not an east–west land bridge but a north–south chain of underwater mountains running from the Arctic Sea almost as far as Antarctica: the Mid-Atlantic Ridge.

By the 1930s, navies' routine use of submarines was ushering in a new age of scientific exploration. In the years leading up to the Second World War, submarine expeditions began to reveal the curious character of the ocean floors. Gravity measurements at sea showed that the ocean was underlain by basalt – far denser than the granite-rich continents. But still no evidence emerged for drowned land bridges.

The war itself had a galvanizing effect. Submarines weren't simply an exploration tool, they were a lethal threat. Heavy maritime losses from German U-boat attacks convinced the Allies that better ocean science offered military advantage. An army of geophysicists were enrolled into Allied naval research. Sonar scanned and probed the ocean floor. Marine magnetic studies tested submarine detection. At the end of the war, scientists attracted by magnetic minesweeping and magnetizing ships returned to academic life fascinated by the earth's own magnetic field.

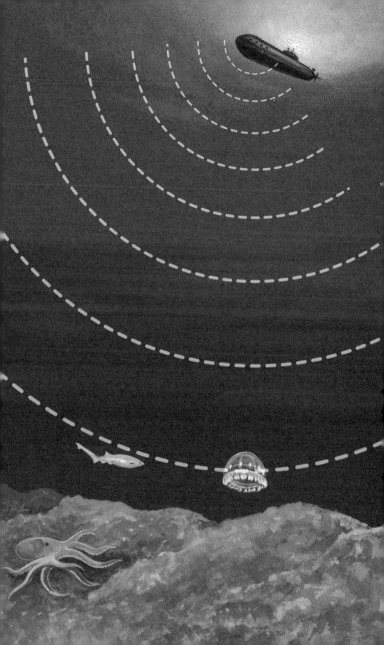

A planetary force field

That our planet acts as a giant magnet had been known since 1600, and by the end of the nineteenth century it was clear it had something to do with its rotation. It turns out that the spin of the globe causes liquid iron in the earth's outer core to slosh about, and that motion generates electrical energy which, in turn, induces a local magnetic field within the planet. The result is we live on a homegrown, self-sustaining dynamo.

The planet's magnetic forces might seem huge but in fact they are hundreds of times weaker than the field between the poles of a toy 'horseshoe' magnet. Indeed, the instruments needed to measure these tiny geomagnetic signals are so sensitive that even a ticking wristwatch upsets them. And yet earth's geomagnetic heart throws a protective shield around the planet. Invisible lines of magnetic force fan out from the South Pole and converge at the North Pole, bowing outward like greatly exaggerated lines of longitude. These force lines intercept streams of high-energy particles ejected from the sun, deflecting them poleward and shedding photons as shimmering light displays – auroras – that streak across polar skies.

It is because earth's magnetic force lines radiate from south to north that the magnetized needles in our compasses also point to north. The arrangement of the force field means that a compass needle at the equator will lie flat, but will become more inclined at increasing latitudes, eventually standing straight up on end at the poles. And just as the unerring compass needle directed the travels of the early navigators, so natural magnetic needles locked into rock set the course for twentieth-century planetary explorers.

The palaeo-magicians

As magma cools and crystallizes, or sediment settles, internal magnetic forces in iron-rich minerals like magnetite align with the magnetic pole. Rock forming today would have those minerals directed northwards, but the earth's magnetic field has reversed itself hundreds of times, flipping from north to south. The timing is erratic and the impetus for this switching polarity isn't understood, but alternating mineral 'compasses' within rock sequences preserve a record of these global reversals.

The palaeomagnetic fabric not only reveals whether the magnetic pole at the time a rock formed was in the north or south, but also what its original latitude was. Rocks laid down at the equator develop horizontal magnetic minerals whereas in rocks formed at high latitudes they are steeply inclined. From these ancient rock compasses, palaeomagnetists found that strata often originated in latitudes very different to the ones they now found themselves in. A study in 1954 reported that 200-million-year-old rocks in England had started out close to the equator. Either the rocks had moved (and the land mass with them) or the earth's whole magnetic field had wandered.

Years of furious debate and intensive measurement followed, but by the late 1950s a consistent answer was emerging: Europe and North America had shared a common palaeomagnetic path until 200 million years ago, but since that time their courses had diverged. That parting of the ways corresponded to the opening of the Atlantic Ocean. It was the first scientific proof of relative movement between land masses. The palaeo-magicians had conjured up the resurrection of continental drift.

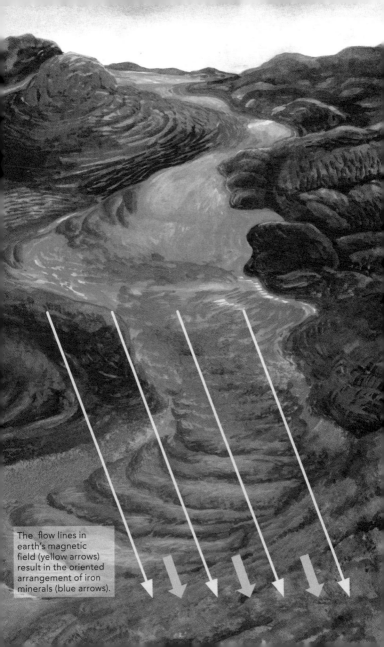

The flow lines in earth's magnetic field (yellow arrows) result in the oriented arrangement of iron minerals (blue arrows).

Southern drifters

In truth, driftist thinking had never disappeared entirely. It had just gone south. After all, it was in the southern hemisphere where the geological evidence for continental coupling was most apparent. So, while most American and European geologists thought they had killed off Wegener's folly, far-flung scientific outposts in South Africa and Australia kept it alive through the 1930s and 1940s. And as geologists down under mapped the land in ever greater detail, more evidence for continental movements emerged.

In New Zealand, geologist Harold Wellman spent the early 1940s mapping a pulverized fracture line along the length of the Alpine mountain chain of South Island. In 1948, he showed that across this Alpine Fault identical rock units had been shunted sideways by 500 kilometres. At the same time, the New Zealand-born Bert Quennell was mapping 107 kilometres of horizontal displacement on the Dead Sea Fault through Jordan, Israel and Syria. What could drive such immense geological shifts?

Working in semi-isolation outside the northern 'fixist' mainstream, the southern 'drifters' were developing their own mechanisms. The most radical came from the Tasmanian geologist Warren Carey, who in 1956 argued that the oceans were relatively recent features on a globe that was increasing in size over time. Carey's 'expanding earth' model conceived the surface of the planet to be the result of it steadily enlarging through the opening of rifts and chasms. Such rifts were known from East Africa and the Red Sea, but might they also extend beneath the oceans?

Bert Quennell mapping the Dead Sea Fault.

Mapping the ocean deep

In the early 1950s, funded by Cold War military operations and the commercial laying of seabed communication cables, two American geologists at the Lamont Geological Observatory near New York began to systematically map the floor of the Atlantic Ocean.

Bruce Heezen was a prodigious gatherer of sea-floor data from countless oceanographic cruises, and Marie Tharp was a 'human computer' converting the raw bathymetric soundings into graphs, profiles and maps. The pair cruised to those parts of the ocean lacking soundings, and by the mid 1950s had amassed an extraordinary data set for the entire Atlantic region. But there was a problem: deep-ocean maps had to be kept classified. It was Tharp who found a creative way around the security restrictions, sketching the numerical data artistically to give a vivid depiction of how the ocean would look if drained of water.

Heezen and Tharp's 1957 'physiographic' geo-artistry of the Atlantic sea floor (and later of the global ocean) was a revelation. Its centrepiece was the vast mid-Atlantic mountain spine, criss-crossed by huge fracture lines and with a narrow V-shaped rift valley snaking along its crest. Heezen and Tharp predicted that this axial rift valley would be traced continuously into the other ocean basins, and over the next few years fellow Lamont oceanographers did just that. By the end of the decade, an almost unknown submarine rift had been mapped for over 37,000 miles (60,000 kilometres) and recognized as the most important structure on the planet. Resembling the stitching of a giant baseball, for Heezen it seemed to perfectly fit an expanding earth.

The geo-poetry of the spreading sea floor

When Princeton geologist Harry Hess skippered an attack transport ship in the Pacific Ocean during the Second World War, he equipped it with a powerful echo sounder and – battle or no battle – never turned it off. His wartime surveys continued into the 1950s, charting an ocean floor rather different to that of the Atlantic: flat-topped submarine mountains rose abruptly from the sea floor while its edges often plunged into deep trenches.

Seamounts and ocean trenches had been known for decades, but in 1960 Hess paired them with ridges in a simple, unified scheme. New sea floor was generated at mid-ocean ridges, fed by upwelling mantle currents in the manner first proposed by Arthur Holmes. Either side of these spreading ridges, conveyor belts of hot, buoyant basalt crust moved apart, progressively cooling and subsiding, volcanic peaks becoming relict seamounts. By the time the sea-floor travelator reached distant trenches, the old, cold and dense ocean crust was dragged down by convection currents and reconverted into mantle.

In Hess's scheme, the planet's ocean floors were impermanent – continuously created at ridges and readily devoured at trenches. Continents – more buoyant and so more permanent – were mobile only because they were passengers catching a ride on a creeping ocean floor. In 1961, Robert Dietz published a similar idea and called it the 'spreading sea-floor theory'. As well as providing a driver for drift, what became dubbed as 'sea-floor spreading' served as an antidote to an expanding earth. The planet itself wasn't spreading – it was maintaining its girth by continually recycling its ocean floor.

Reversals of fortune

In January 1962, Harry Hess travelled to Cambridge University to deliver a talk on the 'impermanence of the ocean floor'. With most geologists still hostile to driftist views, Hess had casually described his idea as 'an essay in geo-poetry'. After all, there was no firm evidence to support the theory of the ocean-floor conveyor belt. But in the audience – entranced by Hess's vision – was a final-year undergraduate student who was determined to change that.

Fred Vine began studying magnetic patterns on the ocean floor. Marine surveys off the Pacific coast of North America in the mid 1950s had revealed an extraordinary set of high and low magnetic 'zebra stripes' running north–south. When Vine's advisor, Drummond Matthews, found a similar magnetic banding on an ocean ridge in the Indian Ocean, the pair combined the ideas of sea-floor spreading and magnetic reversals. They proposed that each time the earth's magnetic field flipped the magma erupting at the mid-ocean ridge inherited the opposite polarity to the previous batch. The result would be black and white stripes of normal and reverse magnetism arranged symmetrically about the ridge.

Unknown to them, a geophysicist in Toronto, Lawrence Morley, had just tried to publish an identical idea in the journal *Nature* but it had been rejected as too speculative. When the Cambridge duo submitted their paper to the same journal a few months later, they had better luck. Published in September 1963, the Vine-Matthews hypothesis eventually became accepted as providing the confirmatory barcode for Hess's sea-floor spreading. Recognizing the twist of scientific fate, a few prefer to call it the Vine-Matthews-Morley hypothesis.

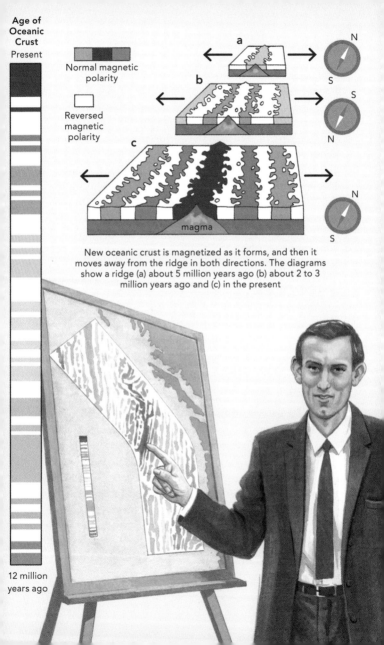

The first touching of a new world

The ocean floor's geomagnetic barcode could only track sea-floor spreading if the ages of the basalt stripes were known. Age-dating the basalt, however, meant drilling in thousands of metres of water and through the thick, muddy seabed. A single deep hole through the entire ocean crust and into the mantle below had been proposed, but in 1961 the ambitious and costly Project Mohole was abandoned after boring just 13 metres into the basalt basement. Reporting from the drilling barge, the writer John Steinbeck celebrated a heroic failure – equivalent to 'Columbus's first feeble voyage of discovery' – predicting that 'on this first touching of a new world the way to discovery lies open'.

Steinbeck was right. The ill-fated Mohole project energized the earth science community. Over the next few years, the USA began to put together the Deep-Sea Drilling Project, to be operated out of Scripps Institute of Oceanography, with a strategy of lots of shallow boreholes rather than grand scientific gestures. A specialist ship was designed and built, a modern version of the venerable *Challenger* a century before.

In 1968, the *Glomar Challenger* set off into South Atlantic waters for its first mission: to put sea-floor spreading to the test. Compelling zebra stripes of magnetic reversals had been revealed across many of the world's ocean ridges but many of the scientists on board remained deeply sceptical; they returned two months later devout converts. The ages from nine short holes drilled across the Mid-Atlantic Ridge perfectly fitted a speculative timescale proposed by the Lamont oceanographer Jim Heirtzler. The oceans now had a barcode reader that could reconstruct their magnetic histories.

Rock, paper, scissors

Fault lines are where much of the planet's geological action is concentrated. Extensional faults dominate the rift valleys that split open the oceans and tug apart the continents. Compressional faults thrust together huge slices of rock to stack up into mountain ranges. And strike-slip faults allow crustal blocks to slide sideways past each other. Generations of geologists were brought up with these three basic classes of fault. Then, in 1965, the University of Toronto geologist Tuzo Wilson discovered a fourth.

Wilson had become fascinated by the great fracture lines that slice through mid-ocean ridges. It had been assumed that these must be giant strike-slip faults laterally dislocating a once continuous chain. But in a moment of insight, Wilson guessed they instead acted as links in the ocean ridge chain. As adjacent but offset ridge segments opened, the sea floor in between was forced in opposite directions. A fault was needed to accommodate the opposing motion but, despite their enormous length on the ocean floor, only the section of fault between the ridges actually needed to move. It was a simple idea, which the showman Wilson loved to demonstrate using only a folded scrap of paper with some cuts.

Tuzo Wilson's 'transform faults' implied that a mid-ocean ridge operated as a simple geometrical system. They also offered a critical test of how sea-floor spreading worked. Almost immediately, Lamont seismologists monitoring a ridge system in the South Pacific took up the challenge. In 1967, it was confirmed that ocean ridge transform faults were moving exactly as Wilson's paper model predicted. But Pacific earthquakes had even more shocks in store.

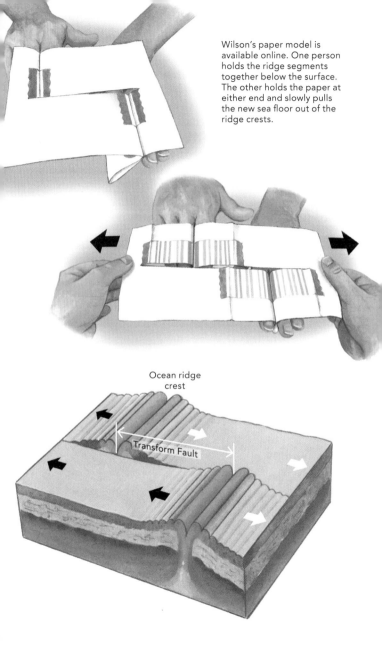

Wilson's paper model is available online. One person holds the ridge segments together below the surface. The other holds the paper at either end and slowly pulls the new sea floor out of the ridge crests.

Ocean ridge crest

Transform Fault

New global shocks

When Eisenhower and Khrushchev began the thorny task of negotiating a nuclear test ban in 1958, the tricky technical issue was how it would be policed. Verifying above-ground explosions was easy, but detecting underground blasts required global surveillance. The fledgling science of seismology was suddenly thrust into the glare of international diplomacy. By 1961, the US government had begun funding a World-Wide Standardized Seismograph Network (WWSSN), spending millions of dollars to install the latest earthquake-monitoring equipment around the globe.

Which explains why, in the mid 1960s, Lamont seismologists Lynn Sykes, Bryan Isacks and Jack Oliver found themselves operating a seismograph network in the Tonga-Fiji area of the South Pacific. Its detailed coverage had verified Wilson's transform faults on nearby ridges but the network also recorded shocks descending 450 miles (720 km) beneath adjacent trenches. 'Dipping zones' of deep earthquakes were known, but the Lamont team's observations imaged a thick, strong slab of Pacific ocean floor being pushed down under the edge of another plate of the earth's crust and actively consumed into the mantle – in what became known as a 'subduction zone'.

Flushed with their success, the trio used the first few years of WWSSN data to reveal the global picture of earthquakes, including the sites of deep tremors. Alongside the first computer-based map of world seismicity, their 1968 paper contained a simple block diagram of the planet's moving parts; both feature in just about every earth science textbook today. They called their paper 'Seismology and the New Global Tectonics', signalling a scientific world teetering on the brink of revolution.

Spinning plates

Computers were becoming essential for making sense of the vast amount of data emerging from the oceans, but a crucial insight would come courtesy of an eighteenth-century Swiss mathematician. Leonhard Euler's 1776 theorem, which describes the motion on a sphere as a rotation about a pole to that sphere, was combined with computer modelling by the Cambridge geophysicist Edward Bullard and co-workers to calculate the 'best fit' of the coastlines of the Americas, Africa and Europe. The result, published in 1965, was compelling confirmation of that centuries-old jigsaw puzzle.

Bullard's graduate student Dan McKenzie recognized that Euler's theorem of rigid body rotations on a sphere described the crustal motions of Lamont's new global tectonics. Over in Princeton, a young geophysicist called Jason Morgan had precisely the same realization. McKenzie got there first. Working with Robert Parker at Scripps, he combined Euler's theorem with computing power to elegantly resolve the crustal motions of the Pacific Ocean. Their *Nature* paper was squeezed out in the last week of 1967, beating Morgan's global synthesis by a few months.

Building on Morgan's work, however, a young French graduate student at Lamont, Xavier Le Pichon, summarized all the relevant data on a map of the world divided into 'plates', and used palaeomagnetic data to calculate their rates of motion. Le Pichon had entitled his 1968 paper 'Sea-Floor Spreading and Continental Drift' but the unifying theory he presented clearly needed a new name. McKenzie and Parker had suggested 'paving stone tectonics'. Thankfully, it would become known instead as 'plate tectonics'.

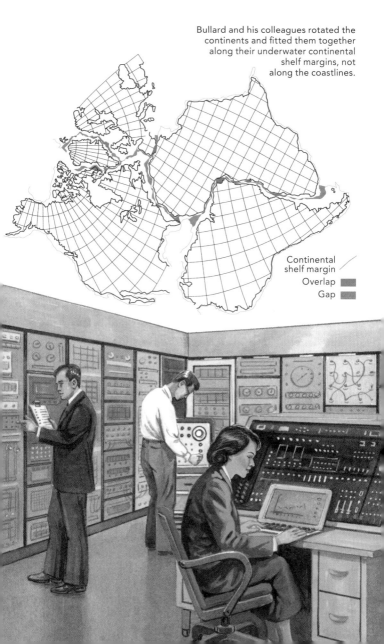

Bullard and his colleagues rotated the continents and fitted them together along their underwater continental shelf margins, not along the coastlines.

Continental shelf margin
Overlap
Gap

Revolution

There is no Nobel Prize for geology, but the Japanese equivalent would later be awarded to Morgan, McKenzie and Le Pichon for 'the initiation of the theory of plate tectonics'. It was their recognition of the rigidity of plates that crucially allowed the surface motions of the planet to be precisely resolved in simple, elegant mathematical terms.

Quite who coined the term itself is vague, but what is clear is that by 1969 the key elements of plate tectonics were essentially in place. During the same few years in which The Beatles emerged, spread and broke up, a cultural revolution had overtaken earth science. Devout fixists became converted almost overnight into zealous mobilists. Much of the revolutionary fervour came from a cadre of mainly young agitators at just a few institutions. It was a tectonic coup largely orchestrated from the university campuses of Cambridge, Toronto and Princeton, the Scripps Institute of Oceanography in San Diego and the Lamont Geological Observatory near New York.

But if the new geophysical theory was to avoid the same fate as Wegener's drift, there were still wrinkles to iron out. There was no clear idea about the forces that drove plate motions, and, more surprising, no direct proof that the planet's surface was measurably on the move. Both these issues would gradually be resolved in the decades that followed. A more acute problem was that the new grand theory of the earth, which had emerged from geophysicists' study of the oceans, had required little input from the continents, or indeed from the geologists that studied them. To secure the revolution, plate tectonics had to work on land too.

Puzzling in continents

While the crustal underlay of earth's oceans is less than 200 million years old, its continents are a puzzling mosaic of geological flotsam and jetsam grafted together over many hundreds of millions to billions of years. Land geologists, aware that the old, messy continents lacked the crisp boundaries and coherent blocks seen by the geophysicists offshore, met plate theory with suspicion and opposition.

A key to unlocking the puzzle of the continents had come from the Canadian geologist Tuzo Wilson. He knew that around North Atlantic shores ancient shallow-marine fossils occurred in two distinct realms. One realm stretched from England and Wales across continental Europe, but also turned up on the eastern seaboard of America, while the other spanned Greenland, Scotland and coastal Norway. Back in 1966, Wilson had proposed that the two realms had been separated by an ancestral ocean, the Iapetus, which had closed along the once continuous line of the Caledonian and Appalachian mountains, prior to the Atlantic opening. Wilson envisaged a global tectonic cycle in which episodes of ocean closure and collision assemble continents only to alternate with periods of break up and sea-floor spreading which undo them.

From the 1970s, geologists slowly began to see plate tectonics in the lay of the land. Buried in many mountain heartlands, they found relict slivers of ancient ocean crust – ophiolites – the telltale vestiges of long-vanished oceans. They gradually recognized that the so-called 'Wilson cycle' could help explain the complex amalgam of the continents. And, after nearly a century of debate, they had finally worked out the origin of mountains: they form when plates collide.

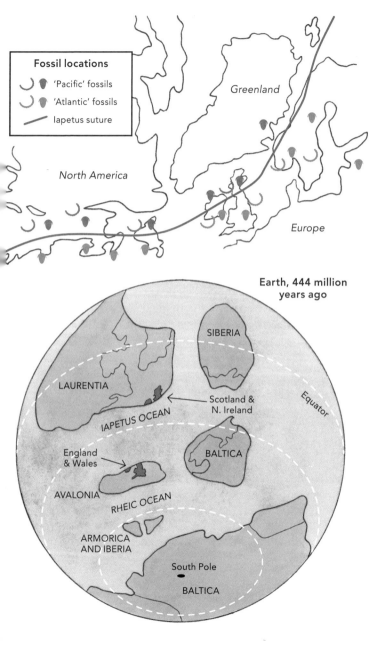

A Cinderella tale

It had taken plate tectonics half a century to emerge from the embers of continental drift, in what one science historian describes as an epic rags-to-riches story: 'The folk tale of Drift is the stuff of myth and legend in which Cinderella, after years of abuse from her vain step-sisters, is visited by the Fairy Geophysicist, is touched by the Magnetic Wand, goes to the Ball and marries the Prince.'

The fifty years following the revolution would not witness such dramatic transformations. The plate tectonic machine was tinkered with and fine-tuned but its rigid cogs remained intact. Moving plates – the most efficient way for our planet to lose its radiogenic heat – involve not only the outer crust but also the strong upper part of the mantle, which together form earth's fractured lithospheric shell. Seven 'major' plates and a scattering of 'micro' plates are continually in motion. Moving a few centimetres a year – roughly the rate at which our fingernails grow – they slide for thousands of kilometres across the weaker, hot mantle below.

In modern parlance, tectonic plates split at 'divergent' plate boundaries, not so much dragged by convecting mantle currents as pulled by their slab edges subducting at 'convergent' plate boundaries, such as those that encircle the Pacific Ocean. Where ocean and continental margins converge, subduction forms volcanic chains and mountain belts, such as along the South American Andes. Plate convergence ultimately brings about head-on continent–continent collision, forming great mountain ramparts like the Alpine-Himalayan chain. Elsewhere, plates grind past each other along 'conservative' boundaries, including the most famous fault on the planet: the San Andreas Fault.

Convergent plate boundary

A fatal attraction

The San Andreas Fault was discovered in 1895 by the Scots-born geologist Andrew Lawson, but it was the devastating 1906 San Francisco earthquake that truly revealed it. Lawson led a forensic survey of picket fences and roads displaced metres across the fault, allowing Harry Reid to develop his 'elastic rebound theory', which confirmed earthquakes as the result of rupture on faults. But why was the fault there?

In 1970, the Scripps geologist Tanya Atwater showed that California was a collage of crustal pieces assembled by plate motion along the San Andreas transform over 30 million years. Today, millions of people live within striking distance of its seismic jolts, but on balance California's great wrench gives more than it takes. Its dramatic tectonic landscape of young oil-rich coastal hills and well-watered central valley support the industries of oil, agriculture, wine and tourism that earn the state billions of dollars each year, many times more than the economic toll of its occasional damaging earthquakes. The human toll, of course, is harder to cost.

Around the world, fault lines have since antiquity offered lush, fertile corridors for human settlement. With the rising population and economic boom of the twentieth century, however, that strategic advantage has turned into a fatal attraction. Villages and towns long tethered to active fault lines have swollen into huge urban targets. Many of the world's largest cities lie in plate boundary zones, and while recent centuries have seen few events like the 1906 earthquake directly strike major population centres, humanity's good fortune is unlikely to last.

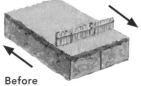
Before
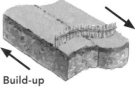
Build-up

Earthquake!
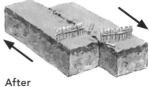
After

Plumes and problems

It was Jason Morgan, the Princeton geologist, who in 1971 recognized that the Hawaiian archipelago is perhaps the most direct expression of earth's moving plates. The islands extend for over 2,000 kilometres, forming a chain that is progressively older from south to north. Morgan argued this marked the track of the Pacific plate over a 'hot spot', where a plume of superheated rock rose from the deep mantle. Leaking up through the thin ocean plate, the Hawaiian plume emerges as outpourings of basalt lava to build the largest volcanic edifices on earth: shield volcanoes.

Mantle plumes may underpin other volcanic islands, such as Reunion, in the Indian Ocean, and Iceland. But it is still unclear how such thermal upwellings – which originate deep down at the core and rise much like buoyant blobs in giant lava lamps – can remain fixed while the mantle they pass through is actively convecting. What is clear is that Hess's notion of spreading ridges being directly fed by convective plumes is wrong. Instead, stretching of ocean crust depressurizes hot mantle material, lowering its melting temperature and transforming it to molten basalt, which then erupts out.

Similarly, the popular view that volcanoes at subduction zones result from the melting of the descending slab below is too simplistic. Instead, water from the wet slab reduces the melting temperature of the mantle above, causing it to partially melt. As that magma rises through the overlying plate, a cargo of crustal contaminants make the melt increasingly viscous, slowing its progress such that it may either 'freeze' at depth, as granitic roots, or reach the surface as a sticky lava loaded with trapped gas, which erupts with explosive violence.

Hawaii

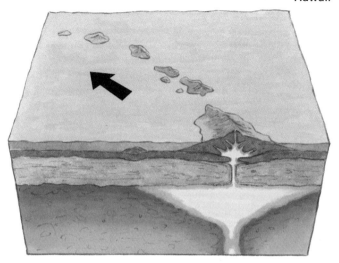

Andes

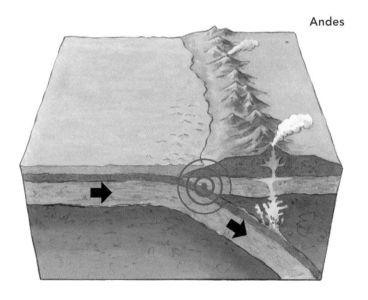

The view from space

In 1957, when the Russian launch of the first satellite, *Sputnik 1*, set the space race running, few scientists considered our planet's surface to be on the move. Although within a decade that mobilist view would become broadly accepted, it would take a further thirty years – and satellites – to actually prove it.

Confirmation would once again come via the tools of global warfare – precise satellite-tracking and gravity-field measurement for the surveillance and targeting of ballistic missiles. In 1986, after five years of space-based monitoring, ground stations in the USA and Norway were found to be moving apart by 2 centimetres per year, while the rate of sea-floor shortening between Hawaii and Tokyo was 8 centimetres per year. The velocities were exactly as predicted from geological rates averaged over millions of years.

The modern mutation from ocean-based to space-borne plate tectonics research has opened an exciting new frontier in the study of our dynamic planet. Satellites such as the new European Sentinel series can detect millimetre-scale changes in earth's surface, imaging the ground motions of individual earthquakes and volcanic crises. Perhaps even more astonishing are the gravity measurements that discriminate the density differences between continents and oceans in unprecedented detail. New gravity images of the ocean floor resolve previously unknown remnants of abandoned continental slivers. Beneath Mauritius and the Seychelles in the Indian Ocean lie the geological roots of ancestral Mauritia and beyond New Zealand is the submerged bulk of greater Zealandia. Not the mythical land bridges but the fractured modern reality of Wegener's lost Pangaea.

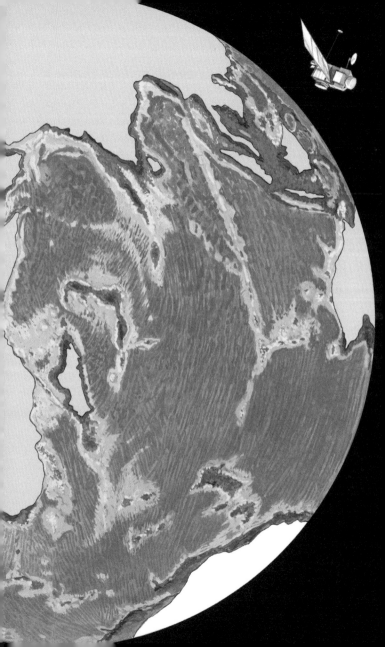

Pale blue dot

Space changed the way we viewed our planet. Earth as a 'pale blue dot' against a spangled blackness became an icon for a new holistic science that emerged in the 1970s. It was led by the English scientist James Lovelock, working with NASA on detecting life on other planets. Lovelock argued that life and its environment on the earth form a single system that self-regulates to maintain a habitable state. His 1979 'Gaia hypothesis' met with an explosion of public popularity and an implosion of scientific rejection.

And yet it was an old idea. Back in 1783, James Hutton – the father of modern geology – had written in his seminal *Theory of the Earth* about the physiology of the planet, asking: 'Is this world to be considered [. . .] merely as a machine . . . [o]r may it not be also considered as an organized body?' A century later, the pre-eminent geologist Eduard Suess coined the term 'biosphere' to describe the 'totality of the animated Earth which live above the lithosphere', renamed the ocean water as the 'hydrosphere' and argued that all earth's spheres (including the atmosphere) were tightly coupled.

Today, the interconnected geological tradition of Hutton, Suess and Lovelock defines modern earth system science. At its heart is plate tectonics. But this encompasses much more than simply the rigid cogs and wheels of a remarkable planetary engine. Plate tectonics regulates the planet, recycling not only its crust but with it water and other ingredients essential for life. Our rocky neighbours – Mercury, Mars, Venus and the moon – may once have been tectonically active but now, lacking moving plates, they are dry and inert. That life exists on that third rock from the sun – which we call 'home' – is thanks to plate tectonics.

Further reading

Peter Molnar *Plate Tectonics: A Very Short Introduction* (Oxford University Press, 2015)

Naomi Oreskes *Plate Tectonics: An Insider's History of the Modern Theory of the Earth* (revised edn; Westview Press, 2003)

Naomi Oreskes *The Rejection of Continental Drift: Theory and Method in American Earth Science* (Oxford University Press, 1999)

Robert Muir Wood *The Dark Side of the Earth* (HarperCollins, 1986)

Mott T. Greene *Alfred Wegener: Science, Exploration, and the Theory of Continental Drift* (Johns Hopkins University Press, 2015)

Henry R. Frankel *The Continental Drift Controversy: Evolution into Plate Tectonics* (Cambridge University Press, 2016)

John McPhee *Annals of the Former World* (Farrar, Straus & Giroux Inc., 2000)

John McPhee *Assembling California* (Josef Weinberger Plays, 1994)

Jack Oliver *Shocks and Rocks: Seismology in the Plate Tectonics Revolution* (Atlantic Books, 1986)

Hazel Rymer and Stephen Drury *Earth's Engine* (5th edn, Open University, 2013)